To . . . .

From . . . .

PIGS
a guide to wonderful wallowing

by James Croft

LAUREL
GLEN

To the uninitiated, pigs have an easy life—with nothing to do all day except lounge around and feed their faces. But behind the scenes a complex and elaborate orchestration is underway: mud for wallowing to the perfect temperature; swill mixed to an exact consistency; tails curled with scientific precision. In fact being a pig is far from simple— it's an art form...

Welcome to the world
of the pig!

Contents

The Basics

gravy

apple cores

newspaper

Swill...

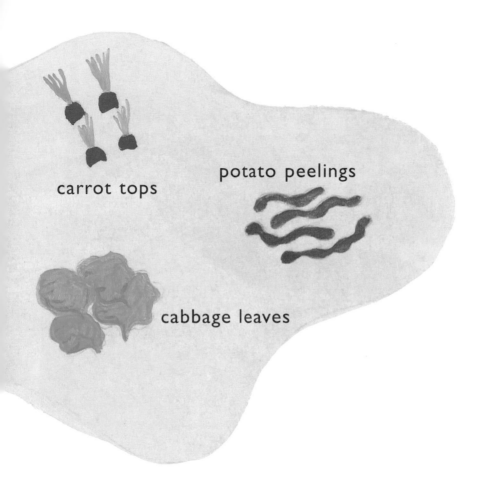

carrot tops

potato peelings

cabbage leaves

...a delicate blend of the finest ingredients

Don't let this
happen to you…

...*do* think ahead

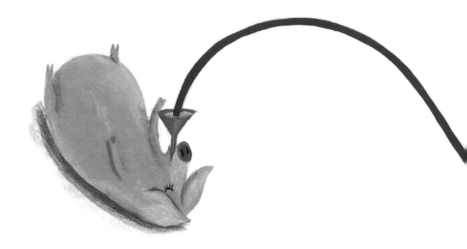

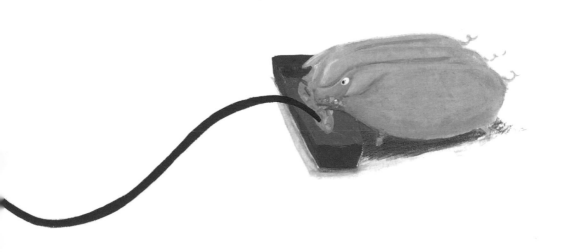

"I love to sleep."

"We love to play."

field mud

Mud types I know:

garden mud

autumn mud

barnyard mud

spring mud

riverbank mud

37°c

98°F

The perfect mud temperature

Wallowing styles:

butterfly

backstroke

My dream mud

empty
packages

pig magazines

It's a pigsty

pig pin-ups

swill chiller

Friends and Foes

My pig pals...

Squealer

Big
Lad

Wilbur

Trotter

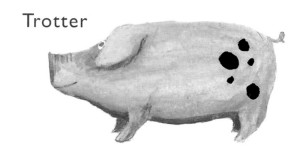

...Rosie with
the piglets

…Sheba Sheepdog is my
best friend

...I get all
the farm
news from
Christopher Crow

...Glenda Goat will eat anything

The chickens are
a bit silly.

Foes:

Foxy

He's rather cunning.

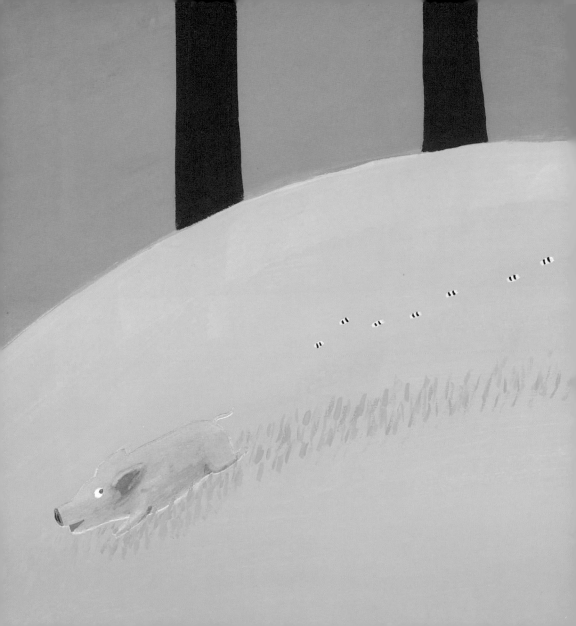

Run away from bees and wasps

But my real
enemy is…

...the Boss Hog

World of Pigs

Curly tails…

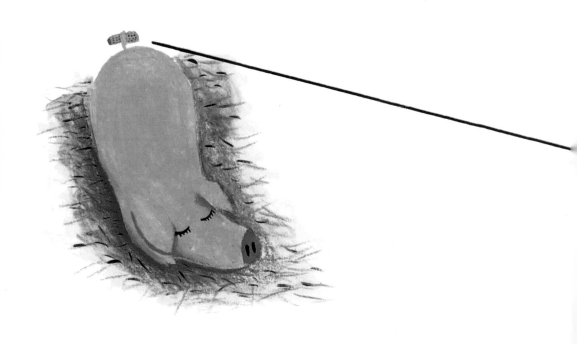

tails in rollers?

There is a more
  drastic way to curl...

Zing!!

hot sauce!

Sty styles...

curl

snake

twist

flick

Sad pigs have...

...straight tails

Pig talk:

...squeals
are for
fright.

This means I'm happy.

Muddy places I love:

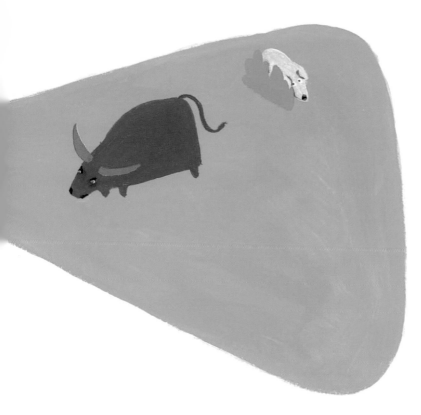

...the bit in the bull's field

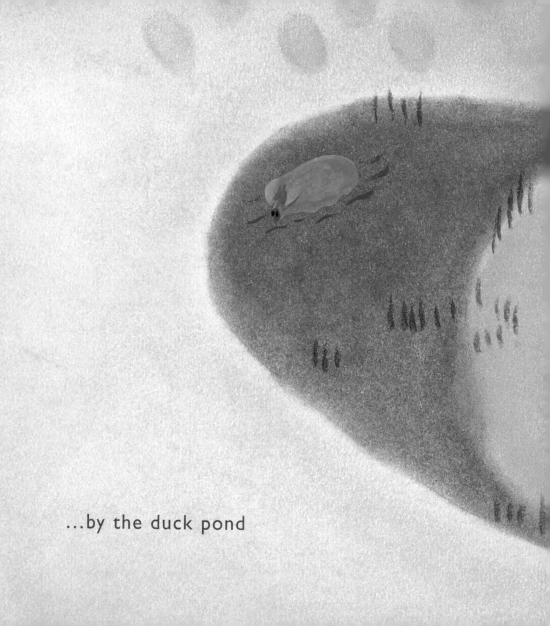

...by the duck pond

...the patch
under the
oak tree

Pigs at Play

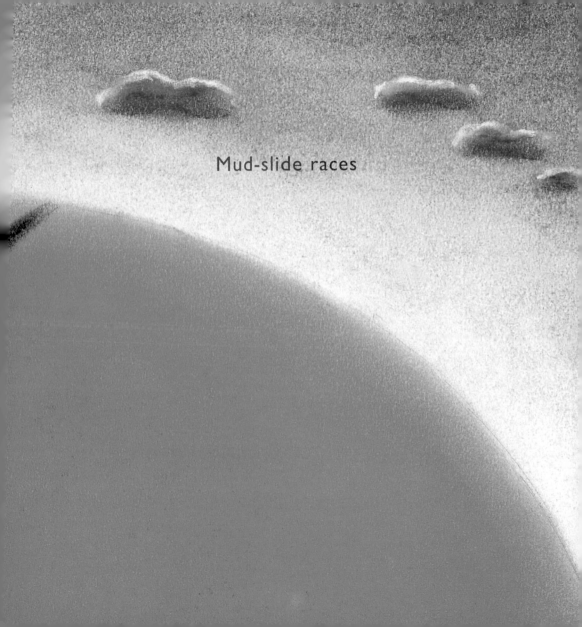

Mud-slide races

Orchard invasions

Nighttime
tractor driving

Out on maneuvers

Barn dances

Mission truffle

Card games

porker faces

Hide and seek

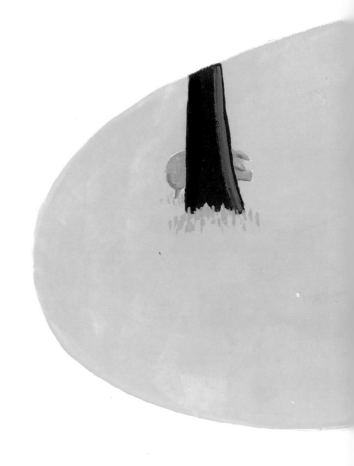

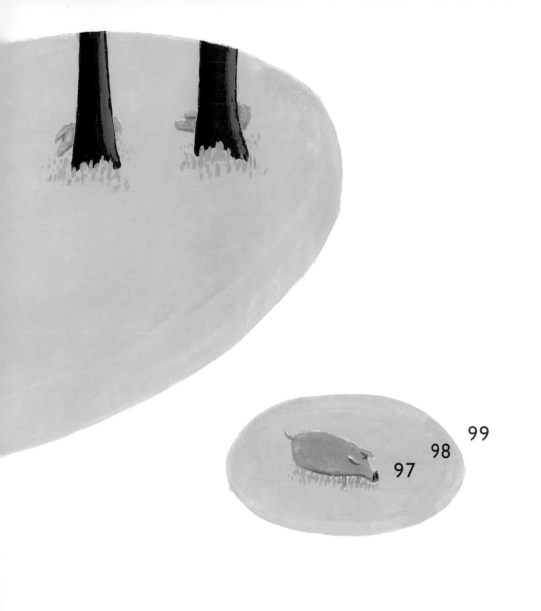

97 98 99

Beauty contests

Farmyard
Fantasies

World-class mud

...Borneo

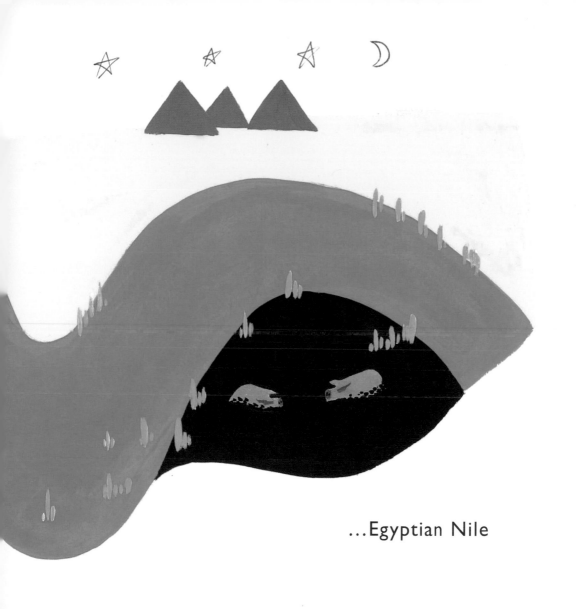

...Egyptian Nile

...Loch Ness

What if pigs could fly?

If pigs had
long legs
they could...

...reach tall trees

...run like gazelles

"You can't catch me!"

If pigs could digs tunnels...

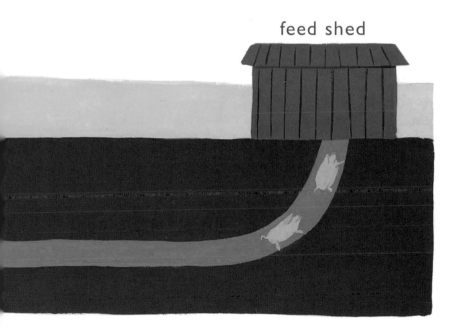

feed shed

...or own a
field of
cabbages

To be
King Pig

But is there anything happier

than a pig in muck?

About The Artist

Born in Yorkshire, James Croft studied in Cleveland, Leeds, and Wolverhampton and now lives in London. Sharing a flat with two college friends, three fish, a cactus, and numerous snails, James is frequently reminded of his rural upbringing which continues to influence his work.

First published in the United States in 2000 by
Laurel Glen Publishing
An imprint of the Advantage Publishers Group
5880 Oberlin Drive
San Diego, CA 92121-4794
www.advantagebooksonline.com

Publisher Allen Orso
Managing Editor JoAnn Padgett
Project Editor Elizabeth McNulty

Author/illustrator inquiries, and questions about permissions and rights should be
addressed to MQ Publications Ltd, 254–258 Goswell Road, London EC1V 7RL;
e-mail: mqp@btinternet.com

ISBN: 1-57145-657-0
Library of Congress Cataloging in Publication Data available upon request.

Printed in Italy

1 2 3 4 5   00 01 02 03 04